LiFi

COMMUNICATION AT THE SPEED OF LIGHT AND THE EMERGENCE OF THE INTERNET OF PEOPLE

GUNTER PAULI
with JURRIAAN KAMP

Praise For The Book

"Congratulations: This books offers a solid base to popularize LiFi and the grand opportunities it offers to society."

—Prof. Dr. Suat Topsu, inventor of LiFi (France)

"LiFi, the exciting technology breakthrough that promises to light the way to a whole new global future, is about to create unimaginable opportunities. Finally we see a chance to bridge the digital gap between the North and the South."

—Ashok Khosla (India) Chairman Development Alternatives and Former Chair of the International Union for the Conservation of Nature

"Deploying LiFi at wide scale worldwide will be yet another a dream come true for humanity; and not the first one to be implemented soon thanks to these brilliant minds."

—Dr. Mariana Bozesan, Ph.D., Dipl.-Inform. (Germany)
AQAL Capital GmbH, Founder & Managing Director
Author of "The Making of a Consciousness Leader in Business"

"In the beginning there was light, LiFI challenges us to return to the beginning and leverage the power of light to sustain our connectivity!"

—Mamphele Ramphela (South Africa)
Former Managing Director of the World Bank,
Former Vice-Chancellor of the University of Cape Town

"When data transmission is no more by radio waves but over light, then we will massively save energy. LiFi presented in this book offers us the starting point to move from the present into the future where we can care for our common house while being connected."

—Rabino Sergio Bergman, Federal Minister of Environment and Sustainable Development of Argentina

"LiFi is big step ahead and opportunity to turn smarter and more efficient our sites and share a unique experience with our sport lovers & users."

—Olivier Colloc-Domancy (France)
Director of Research and Development of Decathlon Mountain Store & Sports

"At last, LiFi is becoming reality. We constantly seem to choose the wrong guy. Tesla would have been a better choice than Edison now we have a chance to make the right choice. LiFi as described by Gunter is the future."

—John Hardy , Founder of the Green School (Bali)

"This book brings a sensational message. Through LIFi it seems possible to secure connectivity through light and thereby significantly reduce energy consumption, eliminate the risk of hacking and limit the power of large companies when it comes to who owns our data."

—The Rt. Hon Anders Wijkman (Sweden)
Co-President of the Club of Rome
Chairman of Climate - KIC

"LiFi is a game-changer: it will humanize the Internet."

—Carlos Moreira , Founder and CEO of WiseKey (Switzerland)

"LiFi presents a great opportunity to speedily and sustainably connect everyone in the developing countries, giving them access to customised information that they require to actively engage in changing their circumstances and bridge the gap that exists today."

—Chido Govera, Founder and Director of the Future of Hope Foundation (Zimbabwe)

"Gunter Pauli, once again, brings transformative technologies to light, literally. For the Pacific Islands, LiFi has the potential to be a revolution to poor connectivity and providing access to Internet to all, while drastically reducing the costs, health risk and islands carbon footprint. LiFi, in the age of the Paris Agreement, must become the way of the future."

—François Martel, Secretary General Pacific Islands Development Forum (Fiji)

"LiFi will contribute to make Internet serve humanity and the common good through faster, safer and healthier communications thanks to light."

—Christopher Wasserman, President Terolab Surface group Founder Zermatt Summit

"Gunter Pauli has been one of the most influential system-entrepreneurs and positive social impact change makers in this century. His immense innovative knowledge, perseverance and creativity brings about the most positive disruptive project that exist today: LiFi."

—Laura Koch (USA) , Chair, Young Presidents Organization (YPO) Social Impact Networks Council

"Gunter's reflection on LiFi brings a new dimension to local action.

Only innovative solutions can respond to the great challenges we face in the 21st century: finding the balance between jobs, environment and the wellbeing of citizens. The City of Roubaix is committed to pioneer to ensure we are part of the economy of tomorrow."

—Guillaume Delbar, Mayor of Roubaix, France

"Trust Gunter Pauli to not be satisfied with the status quo, and to always strive for better. LiFi is much faster and more accurate than WiFi, and uses existing infrastructure (public lighting networks) – opening up a range of new opportunities, of which we've only just scratched the surface. The future is bright, and it's coming to us, quite literally, at the speed of light!"

—Dr Vanessa Tamms
The University of Adelaide (Australia)

"Let there be light: the easiest way to connect people to each other. We can use a simple light ray to access a world of knowledge and sharing."

—Joel Glusman
CEO The Crystal Group (France)

"LiFi is a very innovative technology. It will create a new horizon of communication. Business and society should take note!"

—Yusuke Saraya
President Saraya Co., LTD (Japan)

"This step, which is largely ignored at the present time, may be looked back at as the critical technological infrastructure that opened up a new age of information, flowing between billions of people and a trillion devices."

—Shawn Frayne
Founder and CEO Looking Glass (Hong Kong)
20 Best Brains in the World under 40, Discovery Magazine

LED light and Lifi are "yin and yang". By using the natural capacity to see light, our eyes, we create a most energy efficient light optimiezed to transport information.

—Frans Otten
President Ellipz Lighting (Hong Kong)
Great-grandson of the Founder of Philips

Also by Gunter Pauli

Crusader for the Future: A Portrait of Aurelio Peccei, Founder of the Club of Rome (Pergamon Press, 1987)

Steering Business Toward Sustainability, edited with Fritjof Capra (United Nations University Press, 1995)

Breakthroughs: What Business can Offer Society (1997, Greenleaf Press)

Upsizing: The Road to Zero Emissions (1999, Riemann Verlag)

Out of the Box: 21 ways to be creative and innovative at work (2001)

Zen and the Art of Blue (Commonwealth Press, 2004)

The Blue Economy (USA, Paradigm Press, 2010)

The Blue Economy 2.0 (India, Academic Press, 2014)

The Blue Economy 3.0 (Australia, iXlibris, 2017)

The Third Dimension: 3D Farming and 11 More Unstoppable Trends that are Revolutionizing the Production of Food and Fuel, Regenerating Nature, and Rebuilding Communities (USA, JJK Books www.jjkbooks.com)

Plan A: The Transformation of Argentina's Economy (USA, JJK Books www.jjkbooks.com)

For more information:

www.TheBlueEconomy.org
www.zeri.org
www.GuntersFables.org

For ordering fables by the same author
www.The FableShop.com

For ordering books by the same author
The Third Dimension
www.jjkbooks.com
For ordering The Blue Economy 3.0
https://www.xlibris.com/Bookstore/BookDetail.aspx?BookId=SKU-001116131

The Blue Economy
info@zeri.org

Ordering Information for books in the USA and Canada only
JJK Books
2020 Alameda Padre Serra, Suite 135 Santa Barbara CA 93103
Phone: (415) 595-6451
www.jjkbooks.com

Ordering Information
Quantity sales. Special discounts are available on quantity purchases by corporations, associations, and others. For details, contact JJK Books at the address above. Orders for college textbook/course adoption use. Please contact JJK Books at the address above.

Quantity sales. Special discounts are available on quantity purchases by corporations, associations, and others. For details, contact at the address above. Orders for college textbook/course adoption use. Please contact the address above.

Library of Congress Cataloging-in-Publication Data
ISBN: 978-1-7337177-0-0
First Edition
Interior design: Rick Greer
Production service: Rick Greer

Contents

Foreword

Pioneering communication

In April 1995, Fritjof Capra and I presented our book "Steering Business Towards Sustainability" published by the United University Press per video over broadband internet connecting San Francisco with Tokyo. We were keen on demonstrating that the revolution of the Internet would transform us from citizens to netizens. The book presentation put us down in the history books of the Internet. We consumed 50 percent of the available bandwidth between Japan and the United States. Internet nerds at the time were wondering who were these pretenders pioneering the Cu-SeeMe technology developed by researchers at Cornell University with two powerful Solaris 2 computers purchased from Sun Microsystems. Bill Gates presented his book over the same channel eleven months later.

The experience emboldened Kazuhiko Nishi, the founder of ASCII and a first day partner of Bill Gates at Microsoft, and myself to take the two-way experience to the next level. We decided to organize the first Global Conference on Broadband video internet connecting Nelson Mandela in Pretoria, Jimmy Carter in Atlanta, Shimon Perez in Jerusalem and a gathering of 12 Nobel laureates in Hiroshima under the leadership of Elie Wiesel and coordinated by Ted Koppel, the anchor of ABC News. The first global gathering took place on December 5, 1995 and was reported on the front page of *The New York Times*.

From that early moment, I have witnessed the Internet revolution boom with great interest. However, my primary focus was on the fast track implementation of projects that provide water, food, health, housing, energy and jobs to the people and communities that needed it. The moto was and still is: reach the unreached. I wanted to ensure that people who have hunger and suffer from diseases had a relief that was faster than the connectivity of the Internet. I was traveling the world, learning the science, funding the projects, monitoring the developments, ensuring that we finally reach the unreached. We were regenerating a rainforest in Colombia, farming mushrooms on coffee waste in Zimbabwe and Serbia, and we were learning about producing paper from crushed rocks. The Internet was nothing more than a tool.

That changed in 2007, when I learned during a visit to the Industrial Technology Research Institute in Taipei, about the revolutionary technology of transmission of data and video with light. While the Taiwanese admitted it was not their invention, their reception hall showed this high definition video film projected on a screen without any wires ... the transmission was secured through light. I was fascinated. As an ecologist, I had followed the invention of LED lighting with great interest because it cuts the energy consumption of the incandescent light by 90 percent and the compact fluorescent lamp by 40 percent. I had also seen the projections of sales of LED explode thanks to the industrialization of Dr. Shuji Nakakura's invention of blocking blue light with phosphor to create white light, by leading global players as Toshiba, Nichia, Panasonic, Samsung, Kingsun, Solstice, and Hoyol.

I realized that if we could combine this energy and light revolution with the transmission of data at the speed of light, we were looking at a disruptive technology that could serve the interests of billions of people around the world. I started dreaming: every light bulb could be converted into a high-speed internet router. Soon it was clear that I had to go beyond my dream. There was something better hiding under the logic of digital transmission at the speed of light.

I spoke about this revolutionary breakthrough at the Entrepreneur Summit in Berlin in 2014 where more than 2,000 eager business creators and angels had requested me to lay out the greenest innovations. I mentioned Internet over light. After my talk, I was approached by a representative of the French start-up Thomson Light that had decided to take visual light communication commercial and already had a catalogue with ready to order products. Soon, it became clear to me that these light bulb producers were dependent on the cutting-edge technology of Suat Topsu, professor at the University of Versailles, and the founder and chairman of a French start-up company Oledcomm. As it turned out, Suat Topsu was the inventor of visual light communication using LED. He had avoided attention for nearly a decade focusing instead on solidifying the science behind this breakthrough, and building up hands-on experiences with cities, hospitals and companies.

When I connected with Suat Topsu, I began to see that this new technology followed the core principles of "The Blue Economy" which I had described in my Report to the Club of Rome in 2009. This economic development model proposes that:(1) we use what we have, (2) we generate more value and do not focus blindly on cost, and (3) we respond to the basic needs of all. So, yes, light we have, and the wires to bring the power to the lamps have already been installed. And, if we can indeed create communications at the speed of light—and if time is money— then we can create an incredible platform with innovative services that generate additional value.

Suat pushed my dreams further than I could ever imagine. He demonstrated convincingly that a simple streetlight or a lamp in the office now has the power to convert into a "satellite"—just as the processing capacity of an average car today is the multiple of the computing capacity needed to bring the first man to the moon in 1968. If we can communicate using light that is nearly everywhere, we can re-imagine hundreds, perhaps even thousands of solutions for urgent societal needs. LiFi urgently deserves to become a new platform for technological, social and economic breakthroughs. It has the potential; it is missing awareness. The world of technology and society at large are still amazingly ignorant about the opportunity and those who know have tiny pieces of information that offer no insight into the potential.

That is why this book—true to its message—was written at the speed of light. A cloud of ignorance—blinded by routers, GPS technology and the promise of the coming 5G standard—has to be lifted fast to fully realize the potential of online communication for the benefit of humankind. We are fast approaching a dead-end in the world of computers and communications as we know it.

There is the advent of an incredible merger of technologies: LiFi and LED lamps. There is a breakthrough opportunity to transform the Internet beyond what we have imagined to date. It provides a unique chance to even get our democracy back.

Perhaps the most critical contribution that LiFi offers, is that it allows a shift from the race to bring "big data" under the control of a handful in an "Internet of Things" to design the "Internet of People"; a network for the common good that serves the needs of all. Somehow, the Internet of Things has always seemed a poor perspective for the future. Technology—created by people—, is meant to serve the needs of people, not of things. As we will show in this book LiFi is a tool to strengthen democracy and redistribute power—at the speed of light!

Join me in this quest to transform society. It is easier than we think.

—Gunter Pauli

Zoersel (Belgium), February 2018

Chapter 1

Seeing the light

The LiFi revolution might never have started if Suat Topsu had not listened to his wife.

As an academic researcher and then as a young professor of quantum physics at the University of Versailles, Topsu was involved with autonomous vehicle experiments using the LED lights of cars for inter-vehicle communication as early as 2005. He discovered that cars could "talk" to each other through their headlamps and brake lights.

One day, he came home and his wife, who was pregnant with their third child at the time, told him about a French television documentary on the dangers of radio waves. In response, Sara Topsu decided that the Wi-Fi in their home had to be switched off to protect their new baby's health. "It was a drama for me," Suat recalls. "I needed to work on my computer every day."

Suat initially tried to convince his wife that radio waves were not as dangerous as the documentary had argued. He'd studied a report of the World Health Organization (WHO) on the effects of radio waves on living organisms, based on thousands of scientific studies. Unfortunately, the report stated that pregnant women and their babies were particularly susceptible to prolonged exposure to radio waves. Sara was not amused. "You're a researcher," she told him. "Find another solution."

Installing cables throughout their home seemed a complex,

expensive, and time-consuming solution. Instead, Suat went to his lab to return to his autonomous car research. His scientific mind was stimulated—why wasn't there a better way to communicate with no health risks? He dove into the history of communication. He knew, of course, that Graham Bell had invented the telephone. Not so well-known is that Bell also invented the photophone. In fact, Bell believed the photophone to be his most important invention ever. Of the 18 patents granted in his name, four were for the photophone.

The photophone was similar to a telephone, except that it used modulated light instead of modulated electricity as a means of wireless sound transmission. On April 1, 1880, Bell communicated with a collaborator using a photophone over some 80 meters. A few months later, they bridged a distance of more than 200 meters using plain sunlight as a light source. Bell used a system developed by Samuel Morse four decades earlier when he demonstrated it was possible to communicate over vast distances with light signals. By turning lights on and off, messages could be communicated to a distant observer using Morse code. Bell succeeded in adding an audio channel to Morse's system. Shortly before his death in 1922, Bell said in an interview that the photophone was "the greatest invention I have ever made, greater than the telephone."

Sunlight, however, was not a reliable source for communication. Clouds tended to interfere, and there could be no communication at night. When Bell introduced his photophone, electrical light had only just been introduced in the United States, but that light was not strong enough to transmit information because it wouldn't flicker fast enough. A person using his finger could tap out Morse code much quicker than a light could be switched on and off. This was why the telephone became more popular than the photophone, and electricity and radio waves became the primary channels for communication for the next 100 years.

In his lab, Suat Topsu realized he was able to do what Bell could not in his day. Two technological breakthroughs in light technology allowed for the revenge of the photophone. First was the invention of fiber optics by Corning Glass researchers Robert Maurer, Donald Keck, and Peter Schultz. Fiber optics can carry 65,000 times more information than copper wires. The second invention was the LED (light-emitting diode), in the early 1960s. The first LEDs were low-powered and only produced light in the low, red frequencies of the spectrum. Dr. Shuji Nakamura—professor of material science at the University of California, Santa Barbara—was first to demonstrate the bright blue LED in 1994 in Japan. Blue LEDs led to the development of the first white LED light by adding a phosphor coating, converting the blue light. Nakamura—along with Isamu Akasaki and Hiroshi Amano—was awarded the 2014 Nobel Prize in Physics for the invention of the white LED light.

The LED was the first light that didn't generate heat and did not, therefore, consume a lot of energy. It was cold light produced by a microprocessor, which was able to make the light flicker at high-speed without the need for manual operation of an on and off switch. With it, Suat Topsu realized he could take photonic communications to a new level and that "visible light communication" was the next logical step when adding to Morse, Bell, and Nakamura's research. He saw the emergence of a new information transmission platform, able to change the way we connect to the Internet.

It was possible to multiply the speed of the blinking of light to 100 million times per second (100 MHz), meaning that enormous amounts of information could be transmitted even faster, at the speed of light, which is invisible to the human eye. It would even solve his wife's problem.

Suat Topsu published scientific papers on his invention. Professor Harald Haas, chair of the mobile communications department at the University of Edinburgh picked up his research, having recognized the far-reaching potential of Topsu's discovery. In 2011, Haas gave the TED Talk, "Wireless data from every light bulb," presenting the future of optical wireless communications and introducing the brilliant term, "LiFi." Subsequently, LiFi won a spot on the list of 50 best inventions in *TIME Magazine* in 2011, six years after it had been invented.

In the last few years, Suat Topsu and his Paris-based company, Oledcomm, have introduced LiFi in a museum in Liège, Belgium; in the Paris Métro, having invested 18 months to demonstrate proof of concept; in a hospital in Perpignan, France; and they have introduced experiments in supermarkets. French President Emmanuel Macron has embraced the technology as the spearhead of France's innovation. As Minister of Economics, Macron had already presented the innovation at the World Economic Forum's annual meeting in Davos in 2015.

Still, today, very few people have heard of LiFi. Inventions take time to become popular. Whenever there is breakthrough requiring a fundamental shift, it will inadvertently face ignorance and disbelief, more so than technical, marketing, and legal obstacles. The story of Wi-Fi provides an interesting parallel. The radio wave-based technology was invented in the early 1990s, but it was hardly used for over a decade. It took the arrival of the smartphone in 2007 for Wi-Fi to become the communication medium it is today. The first generation of mobile phones didn't need Wi-Fi because there were no applications requiring it. However, with the arrival of the smartphone, the bandwidth capacity of 3G and even 4G networks soon hit a wall. The only way users could enjoy the full potential of their expensive smartphones was through the additional connectivity of Wi-Fi with which they downloaded audio and video or used to play live games.

Similarly, LiFi will only be used when there is the need for it, and that need is rising rapidly. Faster than we think, even with the arrival of the 5G network in 2020. We are entering a new era that has been coined the "Internet of Things," where all information will be stored in "The

Cloud," rather than on your computer. Today, most communication is still between people—people talking to each other on the phone and people sending texts, emails, and audio messages to other people. It's estimated that a person in the industrialized world will own an average of 7 connected devices in 2025. Most of us will have a computer, a cell phone, a camera, a smart fridge, intelligent keys, intelligent heating and cooling systems, security devices, etc. All of these devices will talk to each other, receive instructions, and act by analyzing the information using artificial intelligence (AI). That is, the "Internet of Things." In your car, you will find a message there is no more milk in your fridge. On your phone, you will turn the heat in your apartment on and off. The coffee machine will start roasting and brewing according to your preferences while you are in the shower. And we are not even talking about the millions of self-driving cars that will find their ways through constant communication over mobile networks. All of that communication requires bandwidth. There are simply not enough frequencies below 10 GHz—the spectrum used for civil communication—to enable all of this data traffic.

The 5G cellphone network that has been in development for over a decade promises broader bandwidths and faster speeds by 2020. It promises to take the present overloaded and clogged system that operates at a maximum speed of 100 Mbits per second to a new online universe at 1 Gigabit per second. An improvement of a factor 10 at the cost of billions of dollars, the installation of thousands of new antennas, and a massive increase in radio wave exposure. Some experts predict the 5G network will be saturated by 2022, two years after it will be launched.

We may wonder whether we need all that communication, whether we want to live in a world of artificial intelligence where devices directly communicate with each other. We argue the need for communication to push innovation, not necessarily to add more gadgets to our modern lives, but to solve the challenges the people and the planet face. It doesn't make sense that necessary innovation is needlessly obstructed by technology. As the laws of physics tell us, nothing moves faster than light, so it makes sense to organize innovation in an environment steered as much as possible by the speed of light.

We need only look at the television or live Internet video to realize how frustrating it is that light travels faster than sound—someone's mouth moves a split second before we hear the voice. Scientists in China have shown that LiFi can provide 252 gigabits/second, 2,500 times faster than the best networks today and 250 times better than 5G. Light-based Internet—fiber optic networks combined with LiFi—will open doors to a whole new dimension of content opportunities and innovation.

Light, including ultraviolet and infrared lights we cannot see, offers a wide spectrum of frequencies. There are a few thousand frequencies available for radio-based communication. As a result, several operators regularly use the same frequencies, which means the baby phone in your home could disturb your neighbor's wireless phone. Providers try to overcome these problems with stronger signals which only invite other, stronger, more energy-intensive signals, leading to a cacophony online and a waste of energy. The full spectrum of light, on the other hand, offers a billion frequencies—one specific frequency for every eight people in the world. In other words, there are no limitations, and we will not run out of capacity if we begin to use light-based communication. That is simple physics.

The laws of physics also guide the concept of triangulation, used to determine a single point in space with the convergence of measurements taken from two other distinct points. Triangulation is the geometrical logic behind modern, satellite-based, Global Positioning System (GPS) location technology. We use more and more satellites in outer space to direct and navigate processes on Earth. The

Internet of Things is increasingly dependent on GPS—soon Google will only enhance its services after you disclose your location. However, the system is hardly precise. Everyone who uses a phone to find their exact location knows that GPS can easily be some 10 meters or more off. That may not matter when you drive from the airport to the city or you try to find a restaurant on a street ten blocks from your hotel, but it becomes a matter of life and death in the case of self-driving cars which have to move forward in narrow lanes only a few meters wide or the use of robots in surgery. We need more precise location services than the current GPS technology can provide. At the First World Congress on LiFi in Paris in February 2018, Industrial groups—like Airbus and Deutsche Telekom—confirmed that LiFi allowed for the necessary location precision, as we shall see in this book.

There is more. There are many important places where Wi-Fi and radio wave transmission does not work, such as under the ground, in mines, or in tunnels. It also does not work when there is a lot of metal around. The port of Antwerp in Belgium is the second busiest container terminal in Europe, processing some ten million containers each year. The port has 2,400 kilometers of railways and two computer server centers that connect some 4,000 kilometers of fiber optic cables to run the complex logistics. The port with all its industries, transportation, and datacenters uses ten percent of Belgium's energy, however, the port's extensive intranet cannot use Wi-Fi. Wireless communication does not work there due to the metal of the containers and the hangars combined with the railroad tracks, high voltage power lines, and industrial complexes, which cause too much interference. This means the containers' QR codes can only be scanned with devices directly wired to the intranet's cables. The Port of Antwerp's headquarters, the emblematic, last building designed by architect Zaha Hadid, suffers from bad Wi-Fi connectivity, as well. Since the architect wanted to preserve the look and feel of the building inside and out, all the Wi-

Fi routers had to be placed in the floor. The stainless steel building that is most impressive to look at is also most problematic from a communications standpoint.

It goes without saying that the efficiency of one of the most modern container ports in the world would be tremendously served by a wireless communications system based on LiFi. That the port has the lights on day and night, operates 24 hours a day, all year round, makes this an ideal site for LiFi. As we shall see in this book, LiFi can vastly improve communications in parts of the economy that cannot be served by Wi-Fi today.

Finally, we need to go back to Sara Topsu, who delivered a healthy third child for the Topsu family in 2009. As the cells in our bodies also communicate through frequencies, there is no doubt that radio waves have an impact on our health. The question is: how big an impact is it? There is a lot of research being done, though the results are not yet completely and convincingly clear for everyone, certainly not for an industry ready to defend billions of dollars of investments. We do, however, know that the World Health Organization (WHO) recommends a maximum exposure to radio wave frequency intensity below 620 milliVolts per meter. This limit is the result of substantial consultation with experts, and there appears to be a clear scientific consensus on this level. In many places—hospitals, offices, coffee shops, and aircrafts—the exposure vastly exceeds the advised WHO norm.

It is tempting to draw a comparison with the tobacco industry of some 40 years ago. At that time, people knew that smoking was bad for you. The question was: how bad? With the impact of radio waves, we find ourselves in a very similar situation. Millennials are the first generation that has grown up amidst today's intensity of radio wave exposure. Do we want to wait for the coming decades to discover the impact of that exposure? Do we want to follow the traditional principle that one is not guilty until proven otherwise, and that we should continue using the technology until it has been proven it is harmful, or do we want to follow the principle of precaution and act like armies, always preparing for a possible attack? Given that our health is at stake, the choice seems to be an easy one. Science needs time to make an unequivocal statement. Society needs the layman's logic that it is better to be safe than to be sorry.

Sara Topsu made a choice that resonates with many mothers. In return, she got more than protection of the health of her baby. She also got an even better and faster Internet connection in her home.

Chapter 3

Locks, screens, airplanes and supply chains

The word "LiFi" was inspired by "Wi-Fi" which was, in turn, inspired by HiFi—or high fidelity. Hi-Fi was the word used to describe the high-quality audio systems introduced in the 1970s. The audio that Hi-Fi provided was very close to original concert sound—the "fidelity" was very high. In the same way, Wi-Fi is wireless communication very close to the original digital communication provided by cables, but the fidelity of Wi-Fi is far from perfect—the Internet is a highly insecure environment, and that is something LiFi will correct.

Despite the complex systems with codes, and response mechanisms with safe keys and firewalls that banks and other companies use to protect your accounts, every system can be hacked. The problem is that Wi-Fi was never designed for security. Rather, it was designed for easy connectivity. Wi-Fi works like a radio: once you tune to the right station or hotspot, you get the music or connection you desire. Both systems are "one way." The radio station does not know who is tuning in and listening. The same applies to hotspots. Everyone can join a hotspot as long as you have the access code or password, which may seem like a major hurdle. If you are an experienced hacker working with a powerful super computer able to run a billion iterations per second, it is not. Radio waves are able to pass through walls so that hackers can work quietly and in their own spaces without being noticed. German chancellor, Angela Merkel, recently found out what this means when it became clear, through Wikileaks, that U.S. intelligence agents had been tapping all conversations on her secured cell phone from the U.S. embassy in Berlin, a few hundred meters from her office.

In a traditional lock system, there is two-way protection: you have a lock and a key, only the key is useless, and the lock without the key is useless, too. Wi-Fi only has a one-way code. Once you are in, you are in. The hacker who published President George W. Bush's paintings had only to think creatively as to which codes the former president might use, based on his personality and personal data. It took only a few weeks to figure it out. Once in, always in, and no one even noticed this hacker had been watching the former president's artistic skills.

Let us compare the current protection of the Internet with the safety deposit box at the bank you may rent to safeguard your jewels. You have a key to your lockbox. That is, you have a key that fits a particular lock. You can only go to the safety deposit box with a bank manager who has another key to open a gate, and then another key to access the second lock on your box. The manager can't get into your box without you, and you can't get in without the manager. It is the double lock that provides tangible security.

LiFi has the potential to completely transform Internet security. We have already seen that the connection only works directly in the light beam. The direct light connection means that a hacker has to be standing next to you. LiFi is not a radio station to which anyone can tune. It can be designed to ensure that only those who should have access can, indeed, have access. This unique feature of LiFi makes it possible to develop perfect double lock software—a key with a lock, just like entering the safety deposit box. This type of security technology has already been developed by WiseKey, a company specializing in Internet security. The double lock technology will be one of the compelling core features of LiFi as it is deployed over the years; it is designed for security.

LiFi security goes further. Each light provides a unique connection point, or ISP address. Once a user connects, the connection is his and his alone. Each LED lamp consists of five, seven, or eleven diodes that jointly produce the light. Each diode can provide a unique connection, giving a connectivity density that, so far, has not been possible. People like to share their experiences at big sports or music events with their friends on social media. So, when Lionel Messi is about to hit a penalty kick for FC Barcelona in a stadium with 100,000 people, many of them have their cameras ready to capture the moment. Subsequently, it turns out to be impossible for all of these fans to upload their pictures at the same time—there simply is not enough bandwidth available. With LiFi transmitting over the thousands of lights in the stadium, there can easily be enough bandwidth for all the fans.

Every user will be able to connect to the next diode with another unique connection. And because there are an estimated 14 billion public lights in the world and each of these lamps will have many diodes, there are billions of connections possible. This is the revolution in the making. Just as there is Moore's Law, there will be a LiFi Connectivity Law. Gordon Moore, co-founder of Fairchild Semiconductor and Intel, argued in a paper in 1965 that the number of transistors in a dense integrated circuit would double every two years. In a similar way, we now predict the number of people connected to the Internet through light will double every year until every light bulb—and then until every diode—is used. So, how long will it take before we are all connected?

Another weak link in the current Internet technology consists of the "connectors." These are the pieces of technology that make the critical links from your router to the communication cable, for instance, but also, the antenna that makes the connection with a hotspot or the service tower of a cell phone company. These essential components have to be of very high quality, as billions of pieces of data run through them. They are also extremely sensitive. A dust particle on a connector between an antenna and a fiber optic line, or a loose connection between wires, means a dropped call or an interruption in service. These connectors need an urgent and fundamental transformation when we begin to transmit data at 10 Gigabits per second. LiFi changes that reality, providing a seamless connection at a tremendous speed because the technology has multiple built-in "connectors." Your phone has front and back cameras that can connect you to a light source. It also has a flashlight that can provide a connection. The backlit screen consists of multiple LED lamps, as well. Each of these lights can provide a connection. Each of these connectors could be transformed into a series of parallel connections to let information flow without interruption. This is what makes LiFi very reliable.

It is important to note that users want speed and bandwidth, they don't want a particular kind of connectivity. Today, when you have no cell phone service, you need to manually select a hotspot or a Bluetooth connection and connect. The upcoming 5G architecture will integrate cellular service, Wi-Fi, Bluetooth, hotspots, and LiFi. The new architecture will automate the selection process. The chip will switch your connection to the best and fastest one available, wherever you are. This can be LiFi, but in a forest on or the beach, it will likely be a cellular connection.

There are environments in which LiFi will beat out all other options, such as in airplanes, for example. More and more airlines offer Wi-Fi onboard their planes. Ideally, airlines would like to offer all of their entertainment through Wi-Fi. There is a simple reason for this. Currently, video screens at passengers' seats are connected to copper wires. This means that there is a lot of heavy wiring under the seats. Airlines want to reduce the weight of aircraft to save fuel, so, replacing existing hardwired connections with Wi-Fi makes sense. At the same time, the

Wi-Fi intensity within the metal walls of a plane will vastly exceed the WHO recommended maximum for exposure to radio waves of 620 millivolts per meter. Moreover, radio waves cause interference with equipment, too, and that is something one would want to avoid on planes.

LiFi eliminates the heavy cables and creates a healthier environment on board. TV, phones, tablets, and phones can all be fed through the individual overhead lights above each seat at speeds no one has experienced so far in the sky. Incidentally, that light also prevents incorrect seating, as in the future, each boarding pass will connect a passenger to the correct, pre-assigned seat. LiFi will provide a very compelling opportunity for airlines.

Warehouses provide yet another great opportunity for the introduction of LiFi. Today, all products in a warehouse—packages at UPS, FedEx, Amazon, or DHL, containers at a terminal, and parts of an assembly line—have bar- or QR-codes that are continuously scanned as products move through the supply chain. The scanning is a vast improvement over the manual supply management of a few decades ago. LiFi, however, will further revolutionize this process. Instead of a QR code that can be scanned, all products will be fitted with a small "I am here" diode which will automatically recognize the products as they move through the beams of LED lights. This will eliminate the need for manual or automated scanning, as it will always be clear as to exactly where any product is. Every single part can be tracked from its point of production throughout its voyage using trucks, rail, or flight, up to its transfer station and warehouse, and then to its final customer or its integration in an assembly. Imagine the number of parts that need to be managed for assembling a plane. Each piece can be tracked each moment, provided there is light. Light-based communication will make logistics infinitely easier, more efficient, and more transparent.

In a very different way, LiFi will support family-building. Many families today deal with the challenge that the very devices parents give their children dominate their relationships. Breakfast, dinners, and other family moments are disturbed by family members checking messages on their devices. MyLiFi presents a great opportunity. A standalone lamp developed by Oledcomm, the company started by Suat Topsu, it offers a LiFi connection but only in its light beams. And the light can be turned off! No light, no connection, meaning more family time. MyLiFi makes the Internet connection local, compared to Wi-Fi which streams intermittently throughout the home. MyLiFi enables parents to control Internet use, a highly desired functionality in many places.

LiFi

Chapter 4

Hackers, gamers and the market entry of a new technology

New technologies face the challenge of entering the market and finding initial acceptance. One can replace a light bulb with a new LiFi-enabled LED lamp, but that is not enough for an online connection. There is the continuing need to develop new software with new capabilities. This is a vast task, the work for which has only just begun. LiFi is an innovation that few people know about so far!

Engineering schools don't offer LiFi-programming courses. The computer expert who comes to your home to update your laptop has never heard of LiFi. When you ask the electrical engineer who fixes a short circuit in your kitchen, he will tell you that he is "not in the Internet business." The architects who design office buildings have lighting teams that have never thought about communication over light. The people who install the cables have always separated electricity and communication wires, but now there is suddenly only one cable. The existing wires end with two different types of sockets at the wall. With LiFi, there is only one wire, but nobody has designed the supply chain for the subsequent integrated socket and plug yet. The hardware we have today can handle maximum data speeds of 100 megabytes per second.

LiFi

LiFi connected to a fiber optic cable will soon handle 252 gigabytes per second, 2,500 times faster than the best wireless connection today. We don't have the chips and computers to accommodate that speed, nor have the connectors been created yet.

When Steve Jobs launched the iPhone, he imagined applications that could be developed by anyone, and before we knew it, over a million apps were available online. In the same way, ever faster Internet over light waves will create a "self-organizing universe." There is no need to plan, and no need to control. Something that one day can hardly be imagined will become indispensable the next day. In other words, LiFi is going to be fun, disruptive, creative, innovative, and game-changing. The question is: who will be the first one to push for adoption and market entry? It seems unlikely it will be the governments, who are always slow to act, nor will it be the existing plethora of online players from search companies to network providers with a vested interested in their current operations. It has to come from a different crowd.

New technology can only successfully enter the market with the support of an initial circle of passionate users. You can't force or push people to accept a new product or service. There has to be an alluring pull driven by curiosity and interest in new opportunities while responding to an innate need that the present communication environment is not able to offer. When Apple began selling the first iPhone, the biggest fans who did not know what, exactly, they were getting, were sleeping in front of the stores overnight. These were highly motivated people, who introduced a new opportunity to the world. They were the happy first users whose enthusiasm spread like wildfire. Can that same buzz be generated for LiFi?

Today, LiFi has no such a group of fans yet. However, there is a group for which the technology opens a highly desirable new dimension: the gamers! An estimated one billion people in the world—one in every eight people—play video games, generating an annual global revenue of over 100 billion dollars. In the United States alone, games are being played in two out of three households.

People are increasingly playing video games online with their peers around the world. It's a global audience of tens of millions of people, located anywhere in the online universe, and guess who has the chance to log the most wins? The one who can make the next move the quickest. In other words, the one with the fastest connection. In Internet gaming, speed is a critical asset. This means the much broader bandwidth and faster speeds LiFi has to offer will have massive appeal for this booming audience.

Moreover, LiFi opens the door of the next dimension in interactive gaming that everyone dreams of, but no one can venture into due to the lack of bandwidth and speed: 3D. Video games in 3D are the cutting-edge for software developers. There is a race to introduce virtual reality in games, but such games cannot be played interactively in cyberspace because Internet speeds cannot accommodate 3D online gamers simultaneously. While the photograph of Messi scoring can wait for an hour or so to be downloaded, in the same time, the gamer would have lost his move. Today, gamers have to settle for oversimplified play compared to what they can imagine and the limits of what their software can create. However, gamers want to play. They will be the ones with the imagination, creativity, and dedication to finding ways to use LiFi.

LiFi will mobilize the ever-growing community of game developers and players who tend to cluster in certain places. In Japan, you will find them in the Akihabara District in Tokyo; in Argentina, they are in Rafaela in the province of Santa Fe; in the United Kingdom, in Nottingham; and the best video game city in the United States is Orlando. These are the places where LiFi will not need to be introduced. The gamers—once aware of the opportunity—will want to be the first to employ LiFi because it will make their gaming the best. It will boost both the content and the scope of games. The gamers are not going to wait for an engineer to show up with an installation package. They won't wait for fiber-optic cables to be equipped with connectors able to handle the new speeds. They will figure things out using what they have because they want to use LiFi now! The gamers will be the pioneers of the implementation of the technology.

There is a second market that is easily accessible to LiFi innovators: the companies investing in new fiber optic networks. Some of these companies are big institutional players with bureaucratic decision-making processes. These are not the prime candidates for the adoption of new technologies. There is, however, an interesting group of smaller fiber-optic network companies who have embraced the opportunity of providing this new service because they expect fiber optics to be the transmission medium of the future, and rightly so. There are cities who have invested in fiber networks for certain neighborhoods or industrial or office quarters. Investments have been made with the purpose of delivering faster Internet speeds in mind, however, with existing technology, the speeds fiber optic networks offer can hardly be enjoyed. The last link is missing. Fiber optic cables may bring the Internet to your home or office at super speeds, but once it gets there, the speed is adjusted to a capacity the WiFi can handle, and that's only a fraction of the fiber optic potential. In other words, these future-oriented investments can be made much more interesting and productive when LiFi service is added. This is why there is a logical and ready audience for LiFi in the operators of fiber optic networks.

Finally, there is a third group of people that may lose their "professions" due to the arrival of LiFi, and that prospect may make them interested in the new opportunities that the technology has to offer. These are the people known as "hackers." As we know, LiFi cannot be hacked. If anyone attempts to come between the light emitting diode and the sender or receiver of a device, the connection will drop. There is only a connection when people use devices in visible light. Hackers are creative and versatile, with an incredible mastery of algorithms. The game of hacking banking information, medical data, and private information for extortion through malware is a known impediment to the Internet. Some of these jobless hackers could be invited to play a role developing the emerging world of a more energy efficient and healthier online universe.

Unfortunately, there is also a new type of danger looming on the Internet. The dramatic increase of speed that LiFi has to offer can also impact the recent troublesome phenomenon of the distribution of fake news. Fake news is based on the capacity of special interest groups to play with algorithms of platforms such as Facebook. These "gamers" of information create news that is obviously false, but they know how to package it in such a way that it is quickly picked up by the algorithms of the platforms forwarding the "news" to the people open to receiving and sharing this kind of information. This distribution is amplified by a few hundred addresses created for this purpose, "senders" who send minor iterations of the original fake news. This wave of false information pushes a separate agenda and confuses with unfounded, purposely created "truths." This new hacking—filling people's minds and space with facts that are aberrations of reality—will benefit from the introduction of LiFi. There is nothing that is all good. Every new technological platform comes with a series of pitfalls and dangers that may also lead to its success.

LiFi will be introduced into the market with bumps and boosts. It will find its way on local levels through logical early adopters. Gamers, hackers, and smaller fiber optic operators can bring this revolutionary technology to local communities with users who will be passionate and determined that this is their preferred choice for connectivity. LiFi's inspiring story will be told through the initial anecdotes and interesting cases that will spread into movements and platforms that will ultimately transform societies.

Chapter 5

Tunnels, tourists and the blind

The story of David and Goliath continues to inspire many. Though Goliath was stronger, David could not have won if he would have played according to Goliath's rules set by Goliath. The same applies to an emerging business coming to the market. Newcomers who don't have a market position have to be creative to find their niches. The traditional competitive analysis following the SWOT model—strengths, weaknesses, opportunities, and threats— does not work for innovative initiatives. It only makes sense when you are readying your company for the fight of the Titans. Start-ups have limited strengths, and they are mostly unknown and unrecognized technologies blended with enthusiasm and passion. Their weaknesses, particularly regarding marketing prowess and capital reserves, are enormous. In that game, the Goliaths will always crush the Davids; the Davids will have to change the rules of the game to succeed.

Soichiro Honda, the Japanese engineer and industrialist who started the Honda Motor Company in 1948, realized this all too well. Initially, Honda produced only motorcycles from a wooden shack. He succeeded because he sold his motorbikes with fuel that was in short supply in Japan after World War II. Honda made his fuel tapping turpentine from the pine trees that covered 70 percent of his country. After successfully selling motorbikes, Honda decided, in 1956, that he would also begin to produce cars. His plan was not well-received, and he should never have succeeded. The Japanese government and the leaders of established car companies—Mitsubishi and Toyota—told him he had no business entering the automobile market.

Honda never did a SWOT analysis. His position was clear: his weaknesses were self-evident and the power of his future competitors overwhelming. The only way to take on the giants of the world was by making a list of weaknesses of the cars of his competitors. That list became so long, he began to develop a car addressing each of the weaknesses of his competitors. He pushed the "go button" in 1956. In 1958, he opened his first assembly plant in Japan. In 1962, he opened his first overseas assembly plant in Belgium while other car makers were debating whether Honda was really making cars or motorized bikes with a car's body. The rest is history, and today, Honda is a successful, fiercely independent multinational.

LiFi will enter the market as a David following Soichiro Honda's strategy. LiFi cannot succeed if it wants to replace Wi-Fi in all cell phones and homes right away and be the preferred tool for connection to today's dominant platforms, from Facebook and Google to Amazon and Microsoft. A starting business should not aim to reach 100 million people to get to the market. Rather, a startup needs to provide proof of concept to demonstrate resilience. This is why LiFi should focus on the needs that cannot be met with the current Wi-Fi and cellular network technologies. The weakness of the old technology is the power of the newcomer. Based on these unique entry points, the new technology can always successfully enter a few market niches. It is a no-risk strategy, a beaten road approach, provided entrepreneurs focus on real needs that have not been met. Once LiFi is nestled into a series of niche markets, the competition will realize that hundreds of these niches make up a mass market, and the domino effect will follow.

Today, the most obvious shortcoming of Wi-Fi connectivity is that it "ends" when we go underground. You can't make calls underground, because there are no antennas powerful enough to pass through the ground, and there is too much metal to cause interference. Radio waves can pass through most walls, but they are blocked by the solid concrete and steel that frames underground structures from parking garages and tunnels to mining shafts and metro stations. GPS doesn't work either. This means that we cannot reach people if there should be an accident in a mine or a terrorist attack in the metro.

LiFi, however, works as long as there is light—every light in a tunnel or mine, together with every hard hat can be equipped with LiFi. In emergency situations, we can know exactly where people are. Mines and tunnels have electricity lines, and LiFi makes it possible to transform these existing power networks into lifelines. It is no surprise that Chile—who has recently struggled with several horrible mining disasters—has been the first country to embrace LiFi technology for its mines. Thousands of kilometers of underground labyrinths require only a change in the lightbulb to offer the communication and the geolocation that meet needs not typically served today.

Metro systems provide the next logical opportunity for LiFi, and Paris is the first city that has begun to implement the technology. RATP, the company that manages the Metro, has agreed to equip all 250 stations with LiFi. One and a half million people use the Paris Metro every day, and they can be guided smoothly through the maze of underground tunnels with nothing more than a cell phone and the thousands of already existing lights to determine the exact location of travelers.

People who are visually impaired may be VIPs, but they have certainly not received VIP treatment when it comes to communications and mobility. At best, there is a converter to braille or an application to translate text to audio using simple directions. It is remarkable that the visually impaired—and their numbers are rapidly rising—have to continuously rely on sticks and guide dogs. But these age-old guiding systems do not work in the complex Metro maze with escalators. LiFi has been guiding the blind through the Paris Metro since February 2018 using a simple app with text-to-speech technology. The light leads the blind through the tunnels, to the escalators, and even to the toilets. This is the groundbreaking change that LiFi offers.

If LiFi can lead the blind, it can also guide tourists who arrive in foreign cities without any local knowledge. Everyone who arrives at Charles de Gaulle Airport in Paris for the first time who does not speak French is as much impaired as the blind. A LiFi-driven app that knows exactly where a first-time visitor is can provide the relevant information at the relevant place, and with simple translation software, in any language. Light is everywhere: from the airport to the subway and the regional trains—everyone can be guided by it and feel as if they are being treated as VIPs. Guiding the blind with light is the first step in the build-up of an infrastructure that will continue to expand until the LiFi city network is fully operational when Paris hosts the Olympic Games in 2024. Once Paris has shown the way, other cities around the world will follow.

LiFi offers a new security dimension to underground transportation systems as well. Terrorists have made public transportation an easy target. Millions of people move through mazes of tunnels, and when the worst happens, no one can communicate. Today, the only viable form of information distribution in these situations is the megaphone. Even emergency exit signs are poor guidance systems to safety, as nobody knows if a sign will send people into the hands of the terrorists. Everyone is familiar with the green emergency "exit" signs in cinemas, hotel corridors, hospitals, schools, sports arenas, et cetera. Worldwide, 80 percent of these signs are made by the French company Legrand. LiFi can turn these signs to red when a particular exit is no longer safe. Legrand sells 500,000 of these systems per year which can be converted into intelligent communication points, sending and receiving data. This means that in emergency situations, through LiFi and LED communication with exit signs and cell phones, people can be guided to the right exits.

LiFi can contribute to another dimension of safety as well. Whatever the reader may think about the use of nuclear power in the rapidly emerging, renewable energy reality, it is a fact of the matter that there are some 450 active nuclear power plants in the world. We know that very dangerous situations can develop when nuclear power plants fail. Japan—and the rest of the world—is still dealing with the aftermath of 2011's Fukushima disaster. In the immediate aftermath of a nuclear crisis, it would be ideal to send in robots—rather than people—to deal with reconnaissance and first remediation. However, robots are useless in the case of nuclear fallout as nuclear radiation distorts the radio wave-based wireless signals that guide them. Seven years after the meltdown in Japan, it is still impossible to enter the core of the disaster zone! Light wave frequencies, however, are not distorted by nuclear radiation. This means that strips of LEDs equipped with LiFi will be able to safely guide robots.

LiFi will change business as well. French corporation Decaux is the number one billboard company in the world. Decaux is increasingly replacing paper billboards with digital ones. Initially, this was a simple efficiency strategy—it's easier to digitally replace ads than to send crews with paper posters through cities—but LiFi offers Decaux an additional, new opportunity. These digital billboards have hundreds of small LED lamps providing backlighting which can turn the billboards into interactive communication instruments. Decaux billboards can offer travelers at a bus stop 3D video downloads in a few seconds, updates on arrivals, suggestions for the connections and how to discover the area while waiting. When this happens, Decaux will no longer be an advertising company—it will enter the communications industry. Their screens will become active, high-speed, communication tools that could soon deliver virtual reality renditions making the vision systems of today look like early day IBM computers: bulky and slow, impractical and expensive.

LiFi becomes irresistible when it adds services to current communication technologies—such as cellphones—that WiFi and satellites cannot provide.

Here is another example. Many people regularly spend frustrating time touring parking garages, searching for that one spot close to the elevators, or that one "green" spot that—according to the sign—should be available on the higher floor. LiFi will transform that experience. As all parking lots have lights in areas where GPS shuts off, drivers that pass the gate and take a ticket can be immediately guided to the one spot reserved exclusively for them. No one else will be guided there, which will save everyone time and frustration. LiFi will also solve the problem of "lost cars." When you enter a parking garage at the airport after a two-week vacation, your phone will know exactly where your car has been parked.

As in Mr. Honda's case, the list of the competition's weaknesses, as well as opportunities for innovative new services, will become longer and longer, and we are only describing the very low hanging fruit here.

Today, all goods in warehouses need to be scanned so they can be located, and cashiers scan the products you buy in the supermarket. In the near future, a little "here I am" diode will be attached to all products and containers, replacing QR codes. This means that all products will automatically communicate with LED lights, so there will be no need to scan anymore.

LiFi will also vastly improve the consumer shopping experience—you will never have to look for a product again. The shopping list stored on your smartphone will have been picked up at the entrance of the store. The LiFi network will quickly calculate distances, check availability, and verify sizes. The system always knows where everything is and offers the best local guidance to take you from one point to the next, securing the shortest distance. The cashier will already "know" the total you need to pay for your groceries the moment your shopping cart passes the checkout.

Research has shown that the installation of LiFi in a supermarket will immediately increase sales by four percent. As it turns out, 60 percent of people leave the supermarket without at least an item or two on their lists they were unable to find. These customers are ready to buy the products, but they cannot locate it in the many aisles of the supermarket. LiFi will solve that problem, increasing consumer satisfaction as well as supermarket sales. Will big supermarket chains be interested in investing in LED lighting with LiFi to increase their sales while also saving on their electricity bills? We don't think this will be a hard investment decision for these chains to make. Companies like InterMarché, Lidl, and E. Leclerc have already made their decisions.

There are many early opportunities, though users may wonder how they will benefit from the LiFi experience before the arrival of the 5G network integrating technology in 2020. All smart phones have cameras designed to capture light on the front and on the back. The cameras are connected to the communication system to send the pictures with technology that is solidly embedded in the phone. There are hardly any phones left that do not have flashlights. This light can flicker a million times per second, and it could be embedded with an encrypted key that must fit into a lock.

Moreover, smartphones are equipped with infrared sensors that adjust screens to be either lighter or darker. Even today, this IR sensor is sufficiently powerful to transfer SMS, WhatsApp, or WeChat messages. The screens themselves consist of LED lights that provide the backlight of the phone. One day, the whole screen can operate like a sender and a receiver at the speed of light. As the compelling opportunities for LiFi arise, early pioneers will find ways to enable the technology with existing hardware.

Chapter 6

Every light turns into a satellite

Almost every new technology requires new infrastructure, new firmware, and new hardware to enter the market. A new software program with many attractive features can often only be run once these three preconditions are met. If Facebook goes from one to two billion users, it will need many more data service centers. An electric car requires a charging station. The arrival of 5G cell phone networks will require newer and stronger antennas, et cetera. These additional requirements often pose the greatest hurdles to the market entry new technologies. The "videotape format war" of the 1980s provides a good example. Most experts agree that the Betamax system was the better technology, however, VHS ultimately became the standard because the technology—led by Sony—offered users much more content in the form of films on video that they could use. In other words, Sony invested in the additional hardware that triggered the creation of the infrastructure (video stores), that led to the winning market entry for the new technology.

Governments have sold licenses for the new 5G networks for billions of dollars. These are investments that very few players can afford. Perhaps the biggest monopoly, and a major obstacle for all mobile connectivity is the Global Positioning System (GPS) navigation technology. All GPS technology depends on American satellites. The United States capital markets, guaranteed by a solid demand from the military, have succeeded in putting some 2,200 satellites in space. Each new satellite requires an investment of anywhere between 50 and 400 million dollars, depending on its use, making them cost-prohibitive for new market players. There are initiatives attempting to save costs with reusable rockets and smaller satellites that cost ten times less, however, paying ten times less for 1,000 satellites is still more money than even the richest economy in the world can afford. This is why the European Union has still not succeeded in uniting forces to successfully launch and operate Galileo, the European satellite program, as a competitive alternative for the American GPS infrastructure.

LiFi provides an extraordinary opportunity against this background, with its ability to offer far more precise geolocation services than GPS without the need for satellites. This is a revolution. LiFi does not need a new infrastructure, as it already has an infrastructure established over the past century: the public light network. There are an estimated 14 billion street lights out there. On top of that, there are one hundred times more light bulbs inside homes and offices. Given all these lights, we have the potential for the most connected society ever—LiFi turns every existing light into a satellite.

Governments worldwide provide public lighting in towns and cities, along streets and roads, and even in shanty towns as an important public service. Light creates a sense of safety and helps to control criminal behavior. The costs of public lighting are never at issue. Governments that need to cut expenses don't do so by turning the lights off in cities.

There are electricity and communication cables along or under every street, in every home and office, under the oceans, and across continents. In other words, the infrastructure required to make LiFi available to everyone already exists and has been paid for. We can use what we already have! This is perhaps the greatest contribution of LiFi technology—there is no need to go through the massive expense of creating a new infrastructure.

LiFi could be distributed through old-fashioned copper electricity cables using current ADSL technology. Data can be transmitted at even higher speeds through new fiber optic cables that can also be used to supply electricity through "power over ethernet" (POE) technology. LiFi can use whatever distribution cable is available, which means governments around the world can introduce LiFi as a public connectivity service—as the backbone for universal access to the Internet. LiFi is a breakthrough in that it integrates high-speed Internet with the public light system. Since these networks are nearly always connected to each home—which has its own local network—the burden of entry is the lowest ever. In other words, it uses an infrastructure that has already been paid for.

Where the European Union failed to establish an alternative for the commercial satellite infrastructure provided by the United States, it is in a very favorable position to take the lead with the introduction of the LiFi technology. Europe has arguably the densest public lighting networks in the world. Instead of investing hundreds of billions to launch satellites, Europe can establish a fast Internet infrastructure, including superior geolocation technology for only a fraction of such an investment. Millions of entrepreneurial ideas are bound to emerge, and the economy will get a boost from the production of the new firmware, hardware, and software.

The introduction of LiFi still requires the changing of light bulbs. Existing lights have to be replaced with LED lamps with drivers and modems. This means a substantial investment but—and here comes the best news—the introduction of LiFi and LEDs lead to major energy savings. There is an immediate power savings of 50 percent with a total savings of 80 percent achieved through intelligent energy management. With LiFi, light and information are managed through the same network which makes it, for instance, possible to dim the street lights after midnight. A mere 20 percent reduction in lumen saves 40 percent in power. And when people leave buildings, the light and data services are immediately shut down.

This massive energy savings makes it possible to finance the conversion to LEDs and LiFi out of existing cash flows with a payback time between three and five years. After that initial period, an ongoing combination of electricity and data services can be offered at 20 percent of today's costs. Still, there is the initial, upfront cost. Even though the money may be recouped within three to five years, changing 100,000 streetlights may cost around 40 million dollars, and there are an estimated 14 billion lights in the world. This means an initial investment of trillions of dollars. Investors always weigh the risks against the returns. The immediate energy savings of LiFi substantially reduces the risks since the repayment of the bill is backed up by energy savings. There is not much of a risk that people will not use the new technology. Cities will not turn off their street lights, and there are few people around today who do not use the Internet. This means the income will be secure.

There is another factor: LiFi turns public lights with essentially no value other than that they are considered an unavoidable and necessary cost into an asset. These same lights now become distribution points for valuable services. A street light being transformed into a "guide" for the blind is a service with value because the European Union requires governments to provide accessibility for visually handicapped people and governments are fined when they do not make their infrastructure accessible for people with handicaps. The absence of this fine can only add to LiFi's value.

LiFi is a standard asset for a source of multiple revenue streams whose value rises over time, which is what investors are looking for. Something that is, today, valued at zero in the books will become a critical component of the information infrastructure—how is that for an investment multiple? An investment in LiFi is solid because it combines the low risk of energy savings with a critical service with for an asset that will only increase in value.

In addition, LiFi investments offer popular carbon credits due to the energy savings generated. The Internet is a critical component of modernization, with the online economy driving modern and emerging societies. The ability to communicate 24/7 driven by Internet and mobile communications is a steadfast growth factor of the economy. It is no surprise the energy consumption of online communication has risen from one to two percent of global energy consumption, and it may even reach six percent by 2030. Every router consumes the equivalent of three good old 60-watt lamps kept on for 24 hours a day. There are buildings with hundreds of routers, and new ones are installed every minute.

LiFi points the way to a cleaner, safer, and cheaper infrastructure that does not need routers, nor does it need expensive satellites in space or antennas every five or ten kilometers. There are trillions of dollars available in pension and impact investment funds in the world. It is hard to think of a more secure, attractive, meaningful investment for institutions that want to contribute to the progress of all. It gives us the confidence to state that any investment in LiFi is a secure investment for the common good with a return that even hedge funds dare not advertise or a risk appreciation that cannot be offered, even by government bonds. In other words, it is too good to be true.

Chapter 7

Healthy hospitals

We know there are certain environments in which we need to be careful with cell phones because they may cause dangerous interference. We know that we should not make calls while getting gas as there is the risk of explosion. We are asked to turn communication devices off when we are on planes as research shows that a concentration of radio waves can distort the migratory patterns of birds, making them lose their age-old routes and end up in places where they have never been before. Other research illustrates the impact of radio waves on the growth patterns of plants. When exposed to intense radio waves, cells do not divide according to healthy, natural, and predictable patterns. In other words, radio waves can impact the behavior of machines, plants, and animals like us.

We also know that the intensity of radio waves continuously increases. In the center of big cities, your phone can connect to dozens of hotspots—the center of London has an average of 65 phone antennas, all of them screaming for attention. More and more devices are able to connect with each other in the "Internet of Things" which leads to a further increase in the intensity of radio waves in smaller and smaller spaces. More and more people have begun to carry two or even three phones. With that intensity, we will experience increased interference and dropped calls and connections. The industry only responds by erecting more antennas for stronger signals which require additional energy, increasing the intensity of the radio waves. We have no idea what this exposure will do to our bodies and health in the long term, which is why it makes sense to apply precautionary principles. It is the same as if we were defending our borders against a possible enemy invasion—we would not wait until you have scientific proof the enemy is about to attack in order to prepare for defense. Instead, we would get ready by doing what is needed, just in case—you would not want to be late.

So far, the research about the impact of all that radiation on human beings is not dramatic enough for the industry to make bold decisions in favor of consumer protection. Western Cartesian logic requires a clear cause and effect and that is—in these situations where there are many different influences—very difficult to determine. The millennials are the first generation growing up in the midst of radio waves in their environment at a modern intensity. We will know much more after a few decades, once they have begun to age. Nevertheless, the recommendation given by the World Health Organization (WHO) after consulting many of the world's leading experts is that we—especially the most vulnerable, like children, the elderly, and the sick—should avoid exposure of radio waves beyond 620 milliVolts per square meter.

It makes sense that sick people in hospitals, whose cells and bodies need to heal, should be exposed to as little radio wave interference as possible. However, as it turns out, it is in hospitals—where more and more equipment and devices communicate via Wi-Fi to better monitor medical processes and maintain equipment inventories—it is not uncommon for radio wave frequencies to measure an intensity of ten times the WHO advisory! There are hundreds of routers in hospitals to connect devices like the intravenous IV drips that distribute fluid and medicine to patients, which have, in many places, replaced old, manual models with autonomous, wirelessly monitored IVs. They are effective, efficient, and precise, and nurses don't have to keep checking the patient, but they also increase the intensity of radio waves in hospitals, and they are at the risk of interference. A brief disturbance in an electrical system can change a critical dose of medicine fed through an automated drip.

Hospitals face a dilemma. More electronic and better-connected equipment allows them to better serve their patients and save on staff that is more and more difficult to contract. At the same time, the wireless communication threatens their very health. There is an urgent need for a better solution. Three years ago, a hospital in Perpignan, France—the Centre Hospitalier—began installing LiFi throughout the entire building. Before the installation, radio wave exposure often measured around 6,000 milliVolts per meter. Today, the hospital measures a maximum intensity of 250 milliVolts per meter in every room, a dramatic reduction, bringing the exposure well below the WHO recommendation of 620 milliVolts per meter. The hospital has eliminated many routers and devices are connected through the LED lights in the corridors and above patient beds. The remaining impact is mostly caused by cell phone antennas in the vicinity whose radio waves keep penetrating the premises.

LiFi enables hospitals to increase efficiency and generate savings. Nurses today, upon switching shifts with co-workers, lose substantial time going from room to room mapping, checking, and monitoring medicine and even patients. If all devices, medicines, and hospital beds have their own little "I am here" diode attached, everything can always be easily traced. The reports that nurses need to write before handing their duties over will be simplified. In addition, visitors can be easily guided to their loved-ones and family members, and nobody will get lost anymore in the extended labyrinth of hospital hallways. LiFi offers the perfect, healthy solution for communication and connectivity within hospitals. To set up a LiFi network in an average hospital would cost about 300,000 dollars, an investment that would be paid back in three to five years out of the energy savings without having to take the operational savings into account.

As an example, the hospital in Perpignan used to have 700 routers in operation, each of which consumed the power of three 60-Watt lamps, 24 hours a day. Every light in the hospital is now an LED light, saving at least 50 percent of the energy while providing communication at the same time. And yes, the lights need to be switched on all the time, however, during the night, the light only needs to shine at a fraction of its intensity, which is enough for the LiFi to work, but not enough to disturb the healthy darkness in the room.

The new LiFi infrastructure offers one more interesting benefit. Today, when doctors prescribe medicines, they use mobile devices that connect wirelessly with the pharmacies at which patients pick up their drugs. This is private and privileged information but somehow, despite regulations clearly prohibiting the sharing of such information, the data is collected through Google et al.'s search systems, albeit anonymously; medical files are worth fortunes to pharmaceutical and insurance companies.

LiFi makes it possible for this communication to be completely private through an intranet connecting the doctor to the patient and the hospital pharmacy without needing a connection to the World Wide Web. Every LED light has a unique ISP-connection, which means the data that is shared is available only to the doctor, the patient, and possibly a nurse on the case. There are no other people that will be connected to the same light, which means that no other people will have access to the information being shared. LiFi makes it possible for patient-doctor relationships to be brought back under the control of the ones who should control it.

That choice does not exist today. Everyone in the hospital is the subject of special data gathering for the sake of information resale.

Trackers can find out what websites are being consulted about cancer research or alternative medicine which can shape the strategy of companies, determine the lobbying efforts of pharmaceuticals that have been blocking insurance to reimburse homeopathy or alternative medicine. As one can imagine, once people become aware, there will be no doubt that demand for this privacy will be explosive.

Whatever time it may take for a scientific consensus to emerge and for the industry to respond, there is no doubt that the majority of expectant mothers would prefer to deliver their babies in a room free of radiation over and above what the WHO has recommended. And which parents would like for their children to spend substantial time in daycare facilities or kindergartens located under large cellphone antennae?

LiFi increases health, privacy, transparency, and efficiency in one of the places where it is needed the most: the hospitals where people spend some of the most vulnerable moments of their lives.

Chapter 8

Closing the gap

In the past 25 years, the Internet has been a major driver of economic development. Societies have been transformed by the online services offered by Amazon, Google, Facebook, Alibaba, Tencent, and a long list of other, online giants. Whole industries—from hotels to taxis—are being overhauled by emerging Internet platforms. Mobile computers—cell phones—with geolocation (GPS) have added a new, online dimension in our daily lives. In short, connectivity is driving growth and progress, remaking society.

There is one challenge: while most people in the Western world speed along the Internet superhighway, half of the world still is not connected. Accessing the Internet requires an infrastructure of satellites, antennas, cables, wires, routers, and servers. These are expensive pieces of equipment that developing countries—struggling with improving the supply of drinking water, providing food security, and offering basic health, education, and infrastructure services—cannot afford. Poor people don't have the cash available to buy subscriptions—or even prepaid cards—to get online access. They can't pay for routers in their homes that cost at least 100 dollars per piece. As a result, we face a situation where half of humanity is engaged in a fast-developing, online, socio-economic expansion while the other half risks falling seriously behind. It is easy to understand what this means for global cohesion. The geopolitical consequences of this trend—from terrorism to migration—regularly make front page news, and the gap is only growing.

There are, however, glimmers of hope. When rice farmers in India are able to cut out the middlemen exploiting them and directly connect with the commodity exchange in Mumbai via a simple cell phone to know the real price on the market, the lives of their families will improve. And, yes, it does help when Facebook launches balloons over rural areas in developing countries to provide connectivity which, of course, includes the caveat that they must use Facebook or face the reality of not having a connection. Still, the impact of these steps pales in comparison with the ongoing explosion of online innovation in other parts of the world.

The bottleneck is the lack of adequate infrastructure. It is hard to see—even after a quarter of a century of Internet use and a decade of smart phones—how the unreached and isolated will catch up without major investments for which there is no money. This will change if the existing infrastructure can be used.

As we saw in chapter five, a major advantage of LiFi is that it can be introduced without a massive prior investment. While it is a fact that 50 percent of the world does not have an Internet infrastructure, 90 percent of the world does have a public light infrastructure. Even where there is no public light yet, it is on top of the development banks' and governments' agendas to provide light to local schools, health care centers, and stores. Through LiFi, we can quickly bring another 40 percent of humanity online using existing public light networks. If it only takes a LiFi-empowered light bulb and a solar panel to bring the rest of the world online, the remaining 10 percent does not need to remain behind for too long. We know that, at night, light creates a sense of security and belonging, and that connectivity changes the livelihoods of people, speeding up economic development, facilitating learning, and building resilience.

This means that a cluster of public street lights, renewable energy, super-effective LEDs, and LiFi technology will make it possible to help close a dangerous socio-economic gap in our world.

The beauty of LiFi is that we can bring connectivity wherever there is light. This light can be a simple light bulb in the village square connected to an old-fashioned copper wire that distributes electricity to a dozen homes. The same wire can be used to carry information using 20-year old ADSL technology. The moment that light bulb is replaced with an LED lamp, the whole square can be connected, which means access to information; to better answers where medical needs are concerned; to better education; to better opportunities for business and trade; and for more sharing within the community. Remember, there are already billions of street lights in the world!

The connection can be made with a simple, second-hand cell phone or computer, and there is no need to buy any cellular services from the multinational providers that make easy money by providing service in developing countries. The free light of the street will replace the expensive subscription to a cellular company and provide a faster, better connection. It is clear that this local street light with LiFi will only connect with the world if it can pick up an antenna, a cable, or a satellite. Often, this is not yet the case. While the system would not currently offer the chance to surf the world, it will offer an introduction to the world through an intranet. The locals can come to a quick agreement as to what information should be available and what must be shared, and this could be posted on a local service so that everyone will have the opportunity to consult.

There is another important dimension: LiFi makes it possible to determine—once again, from the start—which information will be shared and with whom. The World Wide Web grew without a plan, led by the unlimited protection of the freedom of speech. As a result, an estimated 30 percent of today's Internet consists of porn. It is hard to see how that "freedom" serves the common good in society. How would we open up villages and communities to the World Wide Web and provide this kind of accessibility? Most governments that strive to protect the common good do not agree to the simple definition of freedom of speech that has been imposed by a few corporations within the North American legal framework, and they wish to have an additional layer of protection, especially for those who have not previously been exposed.

We are facing the challenge that the Internet as we know it today, through a misplaced focus on the freedom of speech and the emerging phenomenon of fake news, is creating a world that does not match with our reality. Moreover, that reality does not respect the cultural diversity and visions of a healthy, happy, and sustainable future.

It is against this background that the lack of connectivity for half of the world is perhaps a blessing in disguise, as it provides the opportunity to design an Internet strategy and a service portfolio that is a true response to the public's needs.

A few concrete examples might be enough to inspire people to take the initiative to create new businesses based on information that can be easily shared. Instead of pushing for synthetic shampoos and detergents in glittery packaging, anyone interested can learn how to take a kilo of citrus fruit peels, add a liter of water and seven spoons of sugar, and create their own effective and fresh-smelling cleaning products. This information, kept on-hand in each village, could save the skin of millions, the water in the rivers, and avoid the dispersal of waste into the environment littering the place for decades to come.

Similarly, information can be shared about how waste from discarded fruits, vegetables, coffee, and tea can be used to farm mushrooms, providing healthy food while the substrate becomes the ideal feed for chickens; or how the roof of a house or a school could is the ideal spot to farm spirulina, a highly-nutritious algae that needs only sun and warm water to grow. The local production of this would provide all of the trace minerals the children need to grow up healthy. It makes sense that this information would be widely shared. This is what real society is about. This is how we can inspire people to keep cash in the local economy and spur development.

The lists of tested opportunities that have succeeded in other parts of the world and can be shared globally are endless. The new, online network will provide a discovery tour to help develop local economies. Even the installation of the network itself through LiFi-enabled, solar-powered LED street lights, could be explained in detail. And LiFi will grow faster, serving the common good and underpinning local growth with local entrepreneurship.

LiFi could be distributed over light networks which are owned mostly by governments elected to serve the common good. This new information network can function completely separately from the World Wide Web, and as we shall see in chapter 10, out of the reach of Google, Facebook, and their hackers, as well. A local server with all relevant information stored on it will secure accessibility of data that is important to enhance the resilience of the community and the quality of life within it. At the same time, it will assist in learning how to operate and to create a culture of connectivity in phases. This provides a unique opportunity for re-thinking the online information infrastructure we want to create.

We envision that local rural schools and community health centers are the first to connect the surrounding community to an "intranet" providing and exchanging information. This connection will change the way children learn while their parents can benefit from the access to information. Classrooms and homes can be connected to provide a sense of community and better safety. We already know what a simple light bulb powered by a solar panel can do in a village where there was previously no light available after sunset. More reading and more information means more change and better progress. Now, the LED lamp with sun-powered LiFi or the local electricity network can also bring, step-by-step, a connection to the wider world. Information and learning are the beginnings of new possibilities and new habits that can improve and enrich lives. Better ideas as to how to cook food or filter drinking water can travel faster and reach more people, solving more problems. LiFi connectivity can change the world and close the gap. This may very well be the most important contribution of this new technology.

Chapter 9

Safe self-driving cars

Every year, in our world...

...one million people die from malaria;

...half a million people get murdered;

...on average, some 400,000 people die in wars;

...and cars kill 1.2 million people.

The fatality rate of the automobile is equivalent to having ten atomic bombs go off each year. This makes the car, by far, the deadliest "disease" of modern history. We worry about terrorism, but we should really be concerned about the one instrument most of us use every day without even thinking about it. We love our cars because we love our mobility—this is why we take the big risk and high cost involved for granted. We have tightened the rules for drinking and driving, the seat belt is compulsory in the front and back, the airbags give us a sense of protection, the brakes have an anti-locking system, and tires do not spin anymore. A wealth of technologies has been integrated, and yet we do not seem to succeed in making a serious dent in statistics mounted by the killer car.

The problem with cars is that they have human beings at their wheels and human beings make mistakes. This is why mobile robots—driverless cars—could make the world a much safer place. Many people question the safety of the driverless car out of fear the technology may not be reliable, but most of us regularly sit on planes that fly mostly on autopilot. No technology is one hundred percent failure proof, but it will be easy for technology to beat human error, which is why the driverless car will save many, many lives.

We have been told that driverless cars are coming, and fast. Most of us have become familiar with the images of the Google car that relies on very noticeable antennas on its roof which connects to satellites in outer space to provide the car with GPS navigation to avoid causing accidents. And that communication uses radio waves that—as we know—can be hacked. In addition, a radar on the bumper provides a sensing system that works at close range.

It comes as no surprise that governments have begun to express concerns about the arrival of self-driving vehicles, including drones. They fear that autonomous vehicles could be turned into "weapons." In their perspective, any object capable of moving by itself is, by definition, a weapon. Google and the other autonomous car developers will counter that the cars will be programmed not to move when there is a human being in front of them, and the code cannot be changed, which sounds good. However, the hard reality is that radio wave-based systems, including GPS navigation, can be hacked. This means driverless cars could be taken over by someone who wants to use them to cause harm. Yes, technology could help avoid human error, prevent traffic accidents, and make driving safer, but at the same time, technology can never be 100 percent protected against human evil. Trucks have been diverted by thieves who change GPS coordinates to arrange stops at lonely parking lots. Even cargo ships have seen their GPS guidance systems hacked, ensuring that they are located where pirates can act with impunity.

It is interesting to note that the driverless car's future is very much planned in a car-centric-way. The message of the autonomous car is clear: "Here I am. I know where I'm going, and I'm going to avoid accidents. I'm going to follow the white lines, I'm going to stick to my lane, and my car will brake automatically for any danger on the road. And by the way, I will respect all traffic rules." We are planning traffic from the perspective of the individual car. Each car is connected with its surroundings via satellite in outer space. It is this very same, "top-down" approach that characterizes most of the structures and organizations in society. We know there are better and more efficient ways to organize.

If one observes traffic in cities from the sky, the movements of the cars resemble the flows of a colony of ants. We know that nature does not use satellites to organize "traffic," because the ants have other ways to communicate. Flocks of birds rise from a lake without a central command and move together in the same direction due to subtle signaling between the birds which creates the harmonious clapping of the wings, and there are neither collisions nor traffic jams. There is the constant, tacit exchange of information amongst the members of the crowd.

How would it be if cars could "talk" to each other in the same way? Biomimicry is the science that looks at how manmade systems can be inspired by natural patterns. This approach generates more efficient and sustainable frameworks based on systems and logic that have proven their functionality in nature for millions of years. In that respect, it is remarkable that technology giants—like Google and Tesla who have pioneered the development of the driverless car—still focus on GPS navigation and Wi-Fi in their designs. The first test of the viability of LiFi-powered LEDs in vehicles was in 2009 through a cooperation of the University of Versailles and Renault. One would have expected one of these technology leaders to jump on this tremendous opportunity sooner.

LiFi will allow for natural traffic flows that follow natural patterns. It will offer a far more direct and precise communication between cars using their lights, which are critical features of cars. For several decades, many countries have been enforcing the law that car lights should be kept on even in bright daylight. The introduction of LiFi makes it possible for cars to talk to each other with their lights, faster and without the risk of hacking. LiFi car design means integrating all of the car's lights into an internal and external communication network. This will also reduce energy consumption—an important feature for the electric vehicle. LiFi will transform the car into a migratory bird in a flock in constant communication with all other cars around it.

The combination of the lights on the cars with street lights that can be invisibly "on" during the day and strategically placed reflectors will create the perfect light infrastructure for reliable, high-speed communication and guidance systems.

LiFi technology will vastly improve GPS and radar-based systems. Let us start with geolocation. GPS can never turn into a precise positioning tool to within a centimeter of a location. Cell phone tower antennae are at least three kilometers apart, and satellites are at least 20 km above us in space. Moreover, the earth is round, which leads to the Mercator effect: the distortion of a 2D line representing a 3D reality. LiFi will have a huge, positive impact on traffic situations. Radar also helps, but radar communication only warns a vehicle of a problem ahead; radar does not provide communication between the cars in swarm logic.

Here is an example of the difference between LiFi and current GPS-based navigation. If one driver slows down, GPS and radar will tell the next car to slow down as well. Subsequently, the next car will get the same message from the car in front, and the signal will travel this way through the lane of cars on the highway. It is a more efficient system than one with a human driver at the wheel, but it is still a step-by-step response to an incident. LiFi communication is very different: the moment a car has to brake because a deer has crossed the road, that car will communicate instantly—at the speed of light—to all cars behind and around it. All cars will know at the same time what is going on so they can adjust their courses. This is exactly how we see flocks of birds instantly change their patterns. If, on the other hand, the driver only touches the brake and never fully pushes it, then no one behind it will brake, ensuring a continuous flow that would have otherwise been interrupted by the sight of braking lights. LiFi guarantees that all cars waiting at a red light can instantly accelerate at the same time when the light turns green. Imagine how much efficiency this will bring to city traffic!

Because LiFi systems can receive and send all relevant information, they can also help manage the danger of moving vehicles in tunnels in the mountains or under rivers. It is not unusual for the engines or brakes of trucks and buses to overheat in the mountains which can lead to tires catching fire. When this happens in tunnels, it creates a very dangerous situation. Tunnels suffer from a shortage of oxygen, which means that any heat source turns into a charcoal production unit, creating havoc inside at 1,000 degrees Celsius. This is why trucks and buses must have their brakes checked before they drive through long tunnels. Drivers may have to wait for hours to cool down their brakes to ensure the risk is managed by limiting the number of heavy vehicles. It is very easy to install heat sensors in tunnels that can inform truck and bus drivers the exact positions of others through their LiFi connections, and dangerous situations can be prevented.

LiFi brings more efficiency and safety to the emerging world of the driverless cars. In addition, the technology will offer a different and more human way of organizing society. Today, Uber and similar apps show where cars are and how quickly they can help you to get from one place to another. In a LiFi environment, these services can be integrated with information about traffic but also with the needs of people. In emergencies, cars can easily be directed to move people out of dangerous situations. The integration of all information into one network for the common good, will not only reduce fatalities in traffic and increase efficiency, but it will also improve mobility and better serve people's needs.

Chapter 10

The Internet of people

25 years ago, the Internet was an inspiring experiment in democracy. The technology to link computers to communication cables had been developed for military purposes, however, the marriage of the computer and the telephone has turned into a revolutionary new platform of the people, by the people, and for the people. Pioneers everywhere have developed connections and software to link people. If you had a computer and a telephone line you could "dial up" and participate in a conversation around the world. People started to post their research papers, offered their products, and shared their visions and dreams. The World Wide Web was born, a remarkable example of democratic, co-creation.

In the early 1990s, some 1,000 research institutions around the world had connections that were powerful enough to allow for online video conferencing. By 1995, that number had grown to about 10,000. This rapid growth turned into an explosion in 2003, with the introduction of Skype as a platform used by everyone at no additional cost, provided you had an Internet connection. Video conferencing allowed us to add our visual selves to the exchange.

With more and more information—such as text, data, photo, or video—was posted in the new online universe, the need for some organization structure arose. People wanted to easily find the information they needed. The wanted to "search," and early pioneers jumped on the opportunity to offer that service. In 1990, Archie was introduced at McGill University in Canada as the first "search engine." Then came, among others, well-known names such as Alta Vista and Yahoo. In 1998, two entrepreneurs in Silicon Valley founded a company with the strange name of Google.

At first, the search service seemed useful, helpful, and efficient, but then, clever entrepreneurs began to collect data. We should have realized what was going on when Yahoo and Google began to offer free email accounts with abundant storage. In the early Internet days, we bought monthly subscriptions to go online. We had to pay! Suddenly, access to the Internet was free—at least, it seemed like it. We had to pay in a different way: with our data. The search companies turned our data into their revenue. Initially, we were unaware of the process, but our behavior was soon studied and intensely analyzed, and data mining became a science.

The arrival of the smartphone in 2007 accelerated the trend. It allowed the search companies—Google had already dominated the market by then—to follow users with their targeted advertising 16 hours or more a day. They used "algorithms"—a set of rules precisely defining a sequence of operations, according to Wikipedia—to make sure they would personalize content as much as possible. If we liked one story, Google would make sure we would see a similar story to keep our attention and make sure we keep looking at their ads.

The age of distortion had begun. From the "objective" newspaper front pages presenting news selected by a group of editors, we have arrived in a world where the news is presented as a confirmation of our already held views. What we think of as today's news has, in many ways, become Google's selection of the news we have liked, including the always available opportunities to hack—as discussed earlier in this book—election outcomes are now at risk, as we have seen of late.

Meanwhile, Google and platforms like Facebook and Amazon that copied the algorithm magic became outrageously wealthy using our personal information, as well as public information that is not theirs either. Google does not know when an airline flies from New York to Tokyo, nor does the search company know which hotels there are in Tokyo or the best metro line to get from the airport to the hotel. Google also does not know which is the best restaurant close to your hotel. In short, Google hardly knows anything. However, the search engine knows one thing very well: how to find the right information. The company uses a process called "API" (application programming interface) to automatically access information, wherever it is available. Google went to the airlines and said, "Give us access to your schedules, and we will make sure people book more flights with you." They went to hotel chains, towns, and cities in exchange for automatic, unlimited access to schedules, maps, lists, opening times of offices, names of doctors in hospitals, schools that welcomed applications, etc. And Google does not pay for all of this information. It "pays" with something of benefit to all institutions providing information: airlines get more customers, restaurants get more clients, cities get more tourists, and so on. It is a true story, but it is strange that, in the process, Google makes massive money using data that is not theirs. The multinational company does not even pay (sales) taxes in most countries!

A frightening future is rapidly arising in which a handful of companies will control 90 percent or more of all available information. Whenever a new online initiative emerges that shows some success in capturing a part of the market with a new service, it will be quickly bought by one of the giants heavy with cash. The old days of seed money, first and second rounds of investment in a start-up are cut short: millions and even billions will be offered so that everything able to create a dent in the hegemony will be housed under one of these few roofs.

The very World Wide Web that started as a bottom-up people's initiative has become a top-down structure dominated by a few corporate giants. We used to have steel and railway barons, but now we have online kings that are even more powerful. In that respect, the current political debate in the United States about "net neutrality" is of critical importance. There is much at stake here. When lawmakers give in to the wishes of telecom and online corporations and allow them to manipulate the ranking of search results based on which advertiser pays the most, democracy is in peril, and the online experiment in democracy that started 25 years ago will have ultimately failed.

Even if lawmakers do the right thing and protect the public's privacy with regulation, the Internet giants will find their way around them. The European Union offers its citizens privacy and protection of their data, however, everyone who uses Google in the EU "clicks" these rights away when they accept a long screen of "terms and conditions" in small type. The mining of data that belongs to people will continue in the interest of a few billionaires who pay no taxes in the process.

Many people don't care. They are happy with all of the online services, opportunities, information, and music at their fingertips, however, there are growing concerns. Edward Snowden and WikiLeaks have shown that information gathering goes well beyond what a reasonable citizen expects to be necessary to maintain security in society. Increasingly in surveys, people rank privacy as a higher concern than nuclear war.

Against this background, LiFi offers a big opportunity to go back to the drawing board and determine how we want to share what when we connect online. Today, we cannot imagine any other online environment than the World Wide Web. Information "lives" on a website hosted by companies that are a part of or related to the emerging data monopoly, but if we begin to re-imagine online communication, it is quite possible to design a different digital experience. It is possible to rethink Big Data and perhaps even be as ambitious to imagine a democracy free of the meddling of special interest groups operating within the algorithms to win an election.

Most information used and searched by people is local. Where is the hospital? Until what time is that store open? Which Metro line will bring me to the football stadium? Such questions make up 90 percent of all search queries. Why is it necessary to access this kind of information through servers and a global online structure operated and dominated by multinational companies? How would it be if cities brought the information of local services and businesses together in a local information network based on the existing public light infrastructure? Why would local networks not serve local needs and make sure money stays in the community instead of flowing out to Internet companies in other countries?

The answer is, of course, that until now, there was no alternative. The World Wide Web is the only thing we have, and it is good. If it helps local businesses to increase revenue, then why bother? With the emergence of LiFi, the situation changes because there is an alternative. Cities can provide a publicly owned information network based on the existing public light infrastructure, and local software entrepreneurs can use that network to offer local services to local businesses. If in Paris, ask the Parisians which is the best restaurant and if Chef Alain Ducasse wants to promote his three Michelin star place, then let him pay to be ranked first. That money can be made in and stay in the community and taxes may be charged on the amount for added value. There is no deal with Ireland or tax shelter in the Caribbean—the local network only works when you are literally in the light of the city. Your smartphone will connect to the LiFi network as a separate network that is not connected to your regular cell phone service network. You cannot access the LiFi network when you are in another country, but that won't be a problem: you don't need to know which Metro line will take you from the railway station to the football stadium if you are in another city or country.

LiFi

The new LiFi network will be very attractive to local businesses. As we saw before, LiFi's highly precise geolocation services make it possible to direct people much more efficiently than current GPS-based services, and local entrepreneurs who will create the software to open the LiFi network up will make sure they offer better services and at a lower fee than Google. It will not be hard for local stores to discover the advantages of the new kind of network. Imagine, for a moment, that Paris and a few other cities offer LiFi in addition to the existing online web infrastructure. Google should welcome this local competition that will only stimulate citizens to choose, having offered customers opportunities to differentiate.

Moreover, the local LiFi network will be set up with new local government rules to protect your privacy and safeguard your personal data so it will remain yours. If the data can flow from any device over light, there will be no accumulation of Big Data, and it will not take long for this to work with a smart, David-like approach, having changed the rules of the game.

LiFi will open the door to an online renaissance. Based on the public light infrastructure owned by and operated by the local government, local businesses and facilities will support new, local entrepreneurship. This will lead to increased local economic development and money circulating in the local economy. More local connections mean more sharing and more participation, which means a healthier community energized by further democracy. LiFi can revitalize the original online dream.

Chapter 11

The first 100 cities

Throughout history, monopolies have collapsed as new spheres of influence and new technologies emerged. Emperors and kings have experienced that truth. Steel barons and landline phone companies no less. Today, we live in an online communications universe dominated by a handful of giant corporations. When we look into the future, we see the continuation of mergers of technologies. We begin to imagine that the blending of smartphones, virtual reality, blockchain, and artificial intelligence will transform society. And we might think this transformation will be led by the same, online superpowers that already dominate our lives, but that may be a mistake.

A new technology is rising and with it, a void waiting to be filled. The merger of LiFi communications with LED lighting technology will provide a tremendous opportunity for a new wave of bottom-up economic development led by a new breed of entrepreneurs. As we have seen, there is one outstanding feature that marks this new trend: the infrastructure is already available and paid for. There is no need for massive investments in infrastructure to drive this concentration of power. Because the infrastructure already exists, there is an opportunity for a well-known institution—one that is not traditionally expected to drive major change—to play a key role in unleashing the entrepreneurial wave of local economic development.

Most public light infrastructure is owned by cities. In recent decades, cities have lost much appeal as confidence in the government has declined around the world. Today, more people trust companies to deliver the services they require than they expect governments to serve them well. LiFi offers cities the opportunity to re-energize communities, rethink how they might ensure the participation of citizens, and even inspire and involve them. Using the existing public light infrastructure, local services, and local businesses can be connected in a city-owned communications network that better serves citizens and cheaper than the current global advertising- and Big Data-driven World Wide Web.

The start will be easier than we might think. Cities around the world can embrace LiFi and roll out initiatives to save energy, improve health, increase the speed of Internet connectivity, and redefine the concept of Big Data in the service of citizens and the common good.

The system has already been tested. A city can begin—like Paris is already doing—to offer LiFi in an underground metro system. How many metros are in urgent need of changing their lights? How many undergrounds are capable of guiding tourists or visually impaired people through their mazes of tunnels and elevators? Paris has proven it possible to allow blind people to travel through the underground and find a toilet when needed without asking anyone. There are 160 metro systems around the world: New York has the most stations; Beijing the most passengers; and Shanghai the longest lines. If the Metro of Paris has embraced LiFi, one could expect that at least ten percent of the rest of the world will be keen on following this example. This will require an investment of some 500 million dollars.

Cities should take the lead in offering all citizens an exciting, new, online experience. Gamers will declare the cities that lead this their favorite places to live, work and play. Installing LiFi in local hospitals will protect ill people in need of healing from excessive radio waves, which is another logical, early step for any city. The example has been set in a hospital in Perpignan, France. In other words, cities around the world can copy technologies that have already been proven to work in other cities like theirs. So, how many cities are prepared to embark on the hospital conversion? Could we imagine a thousand? The strategy to convert a hospital into one within WHO standards is a "plain vanilla deal"—or maybe a "copy-/paste" situation, as anyone who is keen to learn how it works could ideally go to Perpignan and talk to the pioneers who have already done it.

The challenge will be that there are neither clear standards nor regulations for a new technology like LiFi. Cities are expected to offer public tenders for new projects, but tendering does not work when there are no clear specifications, which is the reason why innovations often do not reach markets fast enough. Another reason is that since there will be technical code written into the process, changing this code will be tough, especially since the experts have no experience in the innovative technology. However, now that the RATP, the company managing the Paris Metro, has taken 18 months to decide on a proof of concept with LiFi and decided to equip all 250-plus subway stations with the technology, why would any other city feel the need to repeat this learning process?

The best way for LiFi to cut through the obstacle of standardization is to focus on delivering the services cities should be delivering but are not, such as guidance for the blind in metro systems. LiFi offers a great solution for that need, which may help overcome administrative challenges.

When services are not being delivered, however, in most cases it may mean that cities do not have a budget for them. The good news is that the investment in the creation of LiFi networks for public service has four sources of revenue. First, a major chunk (if not all) can be paid for out of the energy savings from replacing existing public lighting with LED lamps. This calculation can be done by anyone: consider how many lamps, the cost per lamp, the cost of the installation, the energy savings in kWh, and the rate to be paid to the power company. Still, there is a hurdle to overcome: the initial payment for the installation of the LED lamps implies the need for cash up front with the savings trickling in over the years. It is an investment that is easily earned back, but it begins with the payment of substantial bills. The Paris Metro looked at an initial bill of perhaps 25 million euro. Now, Paris has enough cash reserves and funding to undertake this, but still, if there is another way to get the process moving without impacting the balance sheet, then the decision could be made even easier and quicker.

This is where the 100 Cities LiFi Project comes in. Don't worry: this is not the next global initiative to fail under the heavy burden of administration and bureaucracy.

LiFi is a people's technology that will find acceptance and enthusiasm through their bottom-up, people-driven initiatives to deliver essential community services. But one city is only one city, and even a million euro contract does not mobilize large, solid, and patient capital. The financial challenge will be the same regardless of the city involved. As we have noted, the energy savings of a LiFi-enabled LED infrastructure are up to 50 percent, and with the new services added, it can increase to 80 percent. This makes the installation of LED lamps a priority in the response of (local) governments to climate change. After all, cities are keen to implement the Paris accord.

The European Union recently established a public-private innovation partnership—Climate-KIC (Knowledge and Innovation Community)—to support local governments with the transformation of energy policies to mitigate the impact of global warming. LiFi offers Climate-KIC a concrete platform to serve its mission. Climate-KIC has already started conversations with a dozen cities in Europe about the short-term installation of LiFi networks. Note that this is in the short-term. Installing LiFi networks can be done within months, rather than over years, which leads to a major, additional advantage: the benefits of LiFi can be experienced within a year or so after the initial decision to install a network. This means that local politicians—who are always facing the next election—can show short term results, which will be a powerful driver for LiFi's success. The focus is not on the installation of a technology—the focus is on the provision of services.

This is the second source of financing. Subways have advertising agreements which are likely to continue, as are their busy networks of shops in and around the metro that provide services. The LiFi network permits all of these services online without having to pass through a service provider and without the need for an established search engine because it will go directly over a secure light system. There are a vast array of new services which will be deployed over time, creating an entrepreneurial culture that will transform public transportation as we know it. This will offer multiple revenues for the subways, provided they invest in the server network and software systems.

There is a third source of funding: the increase in asset value. Street lights are, indeed, a cost, and taxpayers are ultimately responsible for the financing. Public transport is also a cost and the investment in subsidies per passenger is high, yet it is funded so people will take the metro, train, or bus rather than clog the roads to the point of collapse. Thanks to LiFi, however, there is new value to be had. Public lighting will become an asset instead of a cost. Public transport will no longer be just a tool to provide mobility; it will also be a unique platform on which to create a community with value-added services. When you have assets that increase in value due to an increase in desired services offered, you will be able to get financing.

Cities brought together through initiatives like Climate-KIC enjoy opportunities to meet the financial challenge. As we have described, LiFi is a solid investment without risk on the technology side and doubt as to performance. It also provides a new platform for revenues. LiFi offers a long-term, stable return backed up by energy savings with lamps that will last a quarter of a century, making it ideal for institutions such as pension and impact investment funds. The risks will be low as the proof of concept has already been established. It becomes even easier when cities team up for joint financing of the investment.

Many big financial institutions do not want to undertake one-off projects about which they know very little. They prefer, instead, to make one decision out of 500 million than two decisions out of 25 million. After all, the process of analysis is the same. Financiers like bigger investments better than smaller ones—they like deal flow: one project after another executed under the same conditions. Instead of one city needing 20 million dollars to install LiFi in a metro and/or hospitals, 100 cities may need two billion dollars, which could very well be an easier investment because the investment fund can now claim savings in major carbon emissions and finances will be realized. It will cut through red tape and reduce the cost of administration; this represents the fourth source of revenues.

LiFi will find its way into the world through a loose global network of cities that will introduce the technology with a clear focus on the common good. The early leaders are already lining up, from Roubaix in France and Kortrijk in Belgium, to Taipei in Taiwan, Sydney in Australia, and Rafaela in Argentina.

The introduction of LiFi networks has allowed local governments to revitalize their communities. The emerging new technology will unleash a new wave of entrepreneurship as cities open a new public network to offer new services. The pioneers may well be the gamers and hackers whose skills have found the common good as their purpose in life and their future professions. Public lighting will become such an important component of the city's services that it will be installed everywhere, making the city more accessible, peaceful, and safe. The city will be more interesting and easier to navigate as information for travelers and consumers will always be readily available. The socio-economic cohesion of a city will increase as the new LiFi network strengthens local businesses and services. Cash will circulate locally.

Local LiFi networks will create and support the community through connecting citizens and local businesses in new and more direct ways. People in LiFi-enabled cities will discover there are other ways to connect than through the online services provided by giant, multinational companies. They will discover they can protect their privacy while enjoying the benefits of online communication. The ultimate message of LiFi is the reawakening of the original online dream of networks operated by the people and for the people in the interest of the common good. This will allow citizens to tackle vast problems that have not yet been resolved. This will permit everyone to participate actively. This will permit us to transform 100 cities at a time.

Welcome, the Internet of the People.

EPILOGUE

This book is only the very beginning, but it sets the stage. It started with the re-discovery of an invention from the 19th century and ended with the view of a horizon beyond anything we can grasp today.

The goal is not to predict the next unicorn—certainly, the strategy is not wealth accumulation. The opportunity is to serve the common good by doing good business with a great new technological platform. The way forward is to inspire city mothers, fathers, and young entrepreneurs to create an environment conducive to serving progress for all. This is the success we have planned for.

A special effort will be made to cluster and pool technologies of all sorts into a hub of innovation that will change the way we communicate. The objective is to communicate and connect with a conscience about the impact on people and the planet with a clarity of the benefits LiFi can bring and an awareness of the opposition this new platform will face.

There is no doubt that LiFi will serve the unreached—from the blind in our society to the orphans in Africa and schools without electricity or the Internet in Latin America. The LiFi opportunity requires an exceptional generation of social entrepreneurs who believe in the common good.

LiFi also carries the potential to change the course of Big Data. We need people who will join to make the change happen, so let us connect and unite our efforts. We have a long walk ahead: it is better to proceed hand in hand.

For further contact and information <pauli@zeri.org>

The Fable of LiFi

Over the past 25 years of engagement in innovations, I have made it a habit to translate each transformative breakthrough into a fable. If we do not inspire our children, then it will be impossible to mobilize the next generation to pursue the work we have initiated but can never complete.

Here is the fable of LiFi, available to all Chinese students as of 2018. Now that the Chinese children know Internet over light is a fact of life, it is our duty to expose children everywhere to imagine this novel Internet service. If their parents do not facilitate it, then the next generation will simply do it!

At the Speed of Light

A family of fireflies gathers in the evening, happily showing off their power to create light. An owl is watching this spectacle.

"Did you know that people are now making light that does not generate any heat?" Owl asks the fireflies. "They call it 'el-ee-dee' light, or LED."

"What is so new about that?" asks a firefly elder. "We have been making cold light from the time we first came into this world."

"Well, that may be true," Owl responds. "But you do not use it to illuminate streets and homes. Now, people even use light to send information."

"So, what is new about that? Before there were radios and phones, people already knew how to send information by using light."

"Yes, but the beauty of it is that now people can have cold light and access to lots of information at the same time."

"Now it is getting interesting," the firefly elder says. "Does this mean that when people go to sleep at night and switch off the lights, they will also switch off the Internet?"

"Absolutely, and that will put a stop to harmful Internet pollution. At least people are now doing more than one thing with what they have," Owl replies.

"This means that people now use light in a way that is cheap, safe and uses very little energy. On top of that, it is also used for sending information. Yes, they are getting smarter all the time."

"Just imagine: a movie of two hours long can be sent to your computer through a ray of light in only thirty seconds!"

"That long? And so fast? Does it not cause a lot of pollution?"

"Pollution is when there is harm caused by having too much of the same. You can have noise pollution when there is too much sound, like making a lot of noise playing the radio and sending out information, but this would be stressful for those who do not want to listen to it. With light, it is hard to have too of it."

"How much light can you see, Mr. Owl?" the firefly asks.

"What a strange question. Of course, I can see only what you can see."

"Not true! Here is a surprising fact for you: there is light that you cannot see, but it can still transmit data."

"Light that I cannot see? That must be very efficient. The best thing about it is that I can go to bed and sleep in peace while there is information swirling around my head at the speed of light!"

About the authors

Gunter Pauli (1956) is an entrepreneur and author who embraces groundbreaking and pioneering initiatives. His latest book, *The Blue Economy*, has been translated into 43 languages and has reached over a million readers. While The Huffington Post named him the Steve Jobs of sustainability, his Latin American friends often refer to him as the Che Guevara of sustainability. The 12 trends in *The Third Dimension* are based on his work with over 200 projects in every corner of the world in the past 25 years. Pauli surfs the waves, intuitively thriving on the transformative and unstoppable trends that no statistics or big data seem to identify

Jurriaan Kamp (1959) left a successful career as South Asia correspondent and chief economics editor at the leading Dutch newspaper, NRC Handelsblad, to found the "solutions journalism" magazine *Ode* which was later renamed *The Intelligent Optimist*. In 2015, Kamp launched a daily online solutions news service, *The Optimist Daily*. Kamp has regularly come in ahead of the curve on stories that advance new visions of our world including a solutions-oriented, optimistic approach of the challenges of sustainability.